Basic Mathematics

Volume-1

NAVIN KUMAR CHOUHAN

Made with ♥ on the Notion Press Platform

www.notionpress.com

This book is Dedicated to My Parents Sri. Ram Pravesh Chouhan and Smt. Mira Devi.

Contents

Foreword

"Welcome to the magical world of mathematics!

As a young learner, you are about to embark on an exciting journey that will help you develop essential skills, build confidence, and unlock the secrets of the world around you.

This book, Basic Mathematics for Standard 1, is designed to make learning mathematics a fun and engaging experience. Through simple language, illustrations, and interactive exercises, we aim to nurture your curiosity and creativity, while laying a strong foundation for future math skills.

In the following pages, you will discover the joys of counting, addition, subtraction, shapes, and patterns. You will learn to solve problems, think critically, and develop a growth mindset that will serve you well throughout your academic journey.

To parents, teachers, and guardians, we invite you to join us on this exciting journey. Encourage your child to explore, experiment, and learn from their mistakes. Celebrate their successes and provide support when needed.

Together, let's make mathematics a delightful and rewarding experience for our young learners!

Happy learning!"

Navin Kumar Chouhan

(M.Com), (CPA-AICPA-Pursuing)

Date 15.04.2025

Preface

Mathematics is the language of numbers, shapes, and patterns. It is an essential tool for problem-solving, critical thinking, and creativity. As a fundamental subject, mathematics plays a vital role in shaping young minds and laying the foundation for future academic success.

This mathematics book for Class 1 is designed to introduce young learners to the fascinating world of numbers and mathematics. The book aims to make learning mathematics a fun and engaging experience, with simple language, colorful illustrations, and interactive exercises.

As an author, I am passionate about making mathematics accessible and enjoyable for all students. With my experience in teaching and writing mathematics books, I have carefully crafted this book to meet the needs of Class 1 students.

About the Author

I, Navin Kumar Chouhan, am a commerce graduate with a Master's degree in Commerce from Indira Gandhi National Open University, New Delhi. I have a strong academic background, having topped my school in the Matriculation Examination.

As a writer, I have authored two books on Business Planning and Basic Mathematics. My professional experience includes working in accounting, taxation, and managerial decision-making for various firms.

Currently, I am pursuing the prestigious certification of CPA-USA (Certified Public Accountant) from the American Institute of Certified Public Accountants.

I hope this book will inspire young learners to develop a love for mathematics and build a strong foundation for future academic success.

Navin Kumar Chouhan

Date : 15.04.2025

Acknowledgments

I am deeply grateful to the individuals who have supported me throughout the creation of this Basic Mathematics book. Their encouragement, guidance, and assistance have been invaluable to me.

First and foremost, I would like to express my heartfelt gratitude to my brother, Mr. Suraj Kumar Chouhan, who has been a constant source of inspiration and support. His suggestions and ideas have helped shape the content of this book.

I am also deeply thankful to my wife, Sabnam Kumari, who has been a pillar of strength throughout this project. Her patience, understanding, and encouragement have helped me stay focused and motivated.

I would like to extend my gratitude to my colleagues who have provided valuable feedback and suggestions on the manuscript.

I am also grateful to my parents, who have instilled in me the value of education and hard work. Their blessings and guidance have been a source of strength and inspiration to me.

I would like to thank everyone who has contributed to the creation of this book. Your support and encouragement mean the world to me.

Navin Kumar Chouhan

1. ADDITION

In previous class, we have learnt basic addition of 1,2 and 3digit Numbers. Now we are going to learn the addition of more than four and mores digits Numbers.

In Mathematics, when we adding two or more numbers together, it's called sum of both numbers, eg- 7 and 4 is equal to 11.

Here, we are going to learn the method and process of Addition with some illustration and examples.

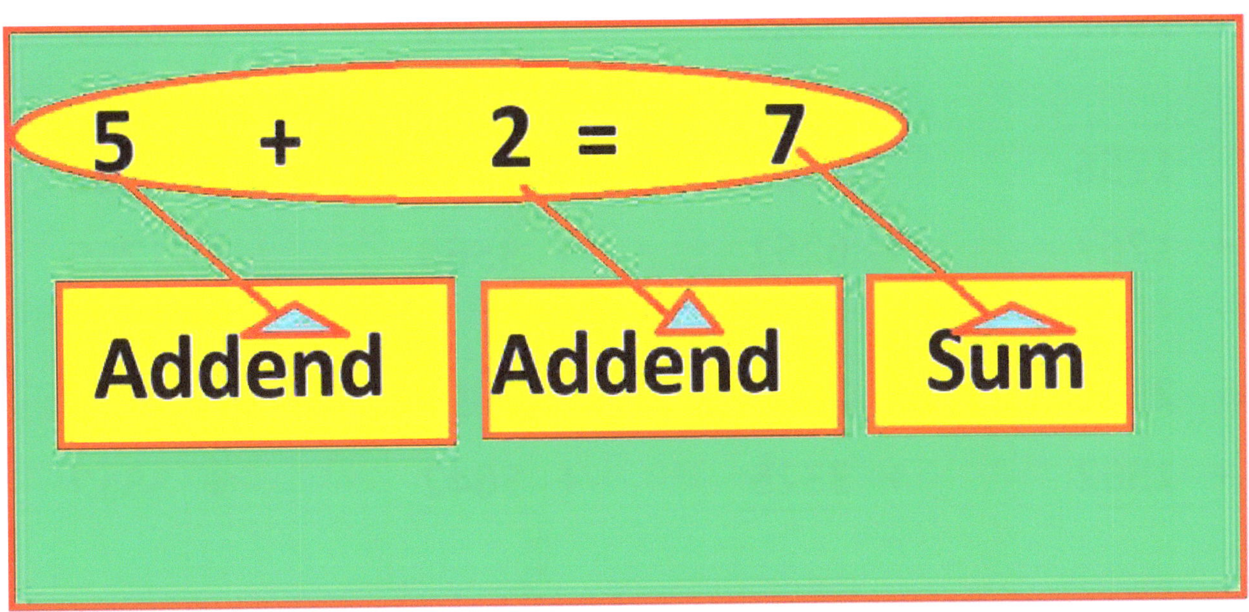


```
  56256
+ 56565
 112821
```

```
  75685
+ 56568
 132253
```

EXERCISE 1.0

5665	2354	5648	7596
+ 9566	+ 2565	+ 8892	+ 1224
6556	5642	9852	6251
+ 1562	+ 3325	+ 5642	+ 3512
1256	1562	1695	1268
+ 1257	+ 6251	+ 3564	+ 1256
2365	2655	2654	2695
+ 1562	+ 3325	+ 5642	+ 3512
3659	3661	3665	3999
+ 9566	+ 2565	+ 8892	+ 1224
4585	4895	4985	4444
+ 1562	+ 3325	+ 5642	+ 3512
5659	5555	5564	5666
+ 1562	+ 1235	+ 5642	+ 3512

EXERCISE 1.1

7845 + 9566 ☐	7777 + 2565 ☐	7888 + 8892 ☐	7864 + 1224 ☐
8565 + 1562 ☐	8889 + 3325 ☐	8444 + 5642 ☐	8665 + 3512 ☐
6556 + 1257 ☐	6542 + 6251 ☐	6444 + 3564 ☐	6664 + 1256 ☐
3500 + 1562 ☐	4000 + 3325 ☐	6000 + 5642 ☐	7001 + 3512 ☐
7001 + 9566 ☐	4005 + 2565 ☐	3005 + 8892 ☐	4550 + 1224 ☐
6005 + 1562 ☐	4855 + 3325 ☐	4000 + 5642 ☐	7000 + 3512 ☐
9500 + 1562 ☐	8500 + 1235 ☐	7500 + 5642 ☐	8900 + 3512 ☐

EXERCISE 1.2

1002 + 9566 ☐	1008 + 2565 ☐	1003 + 8892 ☐	1500 + 1224 ☐
2001 + 1562 ☐	2100 + 3325 ☐	2300 + 5642 ☐	2400 + 3512 ☐
2800 + 1257 ☐	2700 + 6251 ☐	2730 + 3564 ☐	2500 + 1256 ☐
2900 + 1562 ☐	3900 + 3325 ☐	4900 + 5642 ☐	3200 + 3512 ☐
3600 + 9566 ☐	3400 + 2565 ☐	3500 + 8892 ☐	3300 + 1224 ☐
6000 + 1562 ☐	6100 + 3325 ☐	6200 + 5642 ☐	6300 + 3512 ☐
5700 + 1562 ☐	5600 + 1235 ☐	5500 + 5642 ☐	5400 + 3512 ☐

EXERCISE 1.3

56220 + 56156	56222 + 56545	15003 + 35001	15644 + 12245
20015 + 15625	21005 + 33255	23009 + 56425	24009 + 35126
28005 + 12575	27005 + 62515	27305 + 35644	25005 + 12565
29005 + 15625	39005 + 33255	49005 + 56425	32005 + 35125
36005 + 95665	34005 + 25655	35005 + 88925	33006 + 12245
60006 + 15624	61005 + 33254	62004 + 56425	63005 + 35125
57005 + 15625	56005 + 12354	55005 + 56424	54005 + 35125

EXERCISE 1.4

65645	65441	65001	65220
+ 56156	+ 56545	+ 35001	+ 12245
☐	☐	☐	☐
62300	65320	61055	65100
+ 15625	+ 33255	+ 56425	+ 35126
☐	☐	☐	☐
75001	71005	75600	76044
+ 12575	+ 62515	+ 35644	+ 12565
☐	☐	☐	☐
76501	78600	78611	78311
+ 15625	+ 33255	+ 56425	+ 35125
☐	☐	☐	☐
95600	91005	95300	95600
+ 95665	+ 25655	+ 88925	+ 12245
☐	☐	☐	☐
46200	45600	45200	45030
+ 15624	+ 33254	+ 56425	+ 35125
☐	☐	☐	☐
68200	68600	68300	69300
+ 15625	+ 12354	+ 56424	+ 35125
☐	☐	☐	☐

EXERCISE 1.5

6212 6564 + 5656 ☐	5646 6544 + 5654 ☐	6151 6500 + 3500 ☐
4520 6230 + 1562 ☐	6931 6532 + 3325 ☐	6931 6105 + 5642 ☐
6351 7500 + 1257 ☐	6961 7100 + 6251 ☐	6315 7560 + 3564 ☐
6351 7650 + 1562 ☐	7592 7860 + 3325 ☐	6521 7861 + 5642 ☐
5621 9560 + 9566 ☐	5612 9100 + 2565 ☐	1265 9530 + 8892 ☐
6256 4620 + 1562 ☐	6521 4560 + 3325 ☐	6510 4520 + 5642 ☐

EXERCISE 1.6

56461 65441 + 56545 ☐	61512 65001 + 35001 ☐	36512 65220 + 12245 ☐
69310 65320 + 33255 ☐	69310 61055 + 56425 ☐	69310 65100 + 35126 ☐
69610 71005 + 62515 ☐	63158 75600 + 35644 ☐	69150 76044 + 12565 ☐
75920 78600 + 33255 ☐	65210 78611 + 56425 ☐	56100 78311 + 35125 ☐
56122 91005 + 25655 ☐	12652 95300 + 88925 ☐	45611 95600 + 12245 ☐
65210 45600 + 33254 ☐	65100 45200 + 56425 ☐	65810 45030 + 35125 ☐
56845 68600 + 12354 ☐	65891 68300 + 56424 ☐	68541 69300 + 35125 ☐

2. SUBSTRACTION

In previous class, we have learnt basic subtraction of 1,2 and 3digit Numbers. Now we are going to learn the subtraction of more than four and mores digits Numbers.

Subtraction is a mathematical operation that involves finding the difference between two numbers. It's denoted by a minus sign (-). For example:

5 - 3 = 2

Here, 5 is the minuend (the number being subtracted from), 3 is the subtrahend (the number being subtracted), and 2 is the difference (the result).

Here, we are going to learn the method and process of subtraction with some illustration and examples.

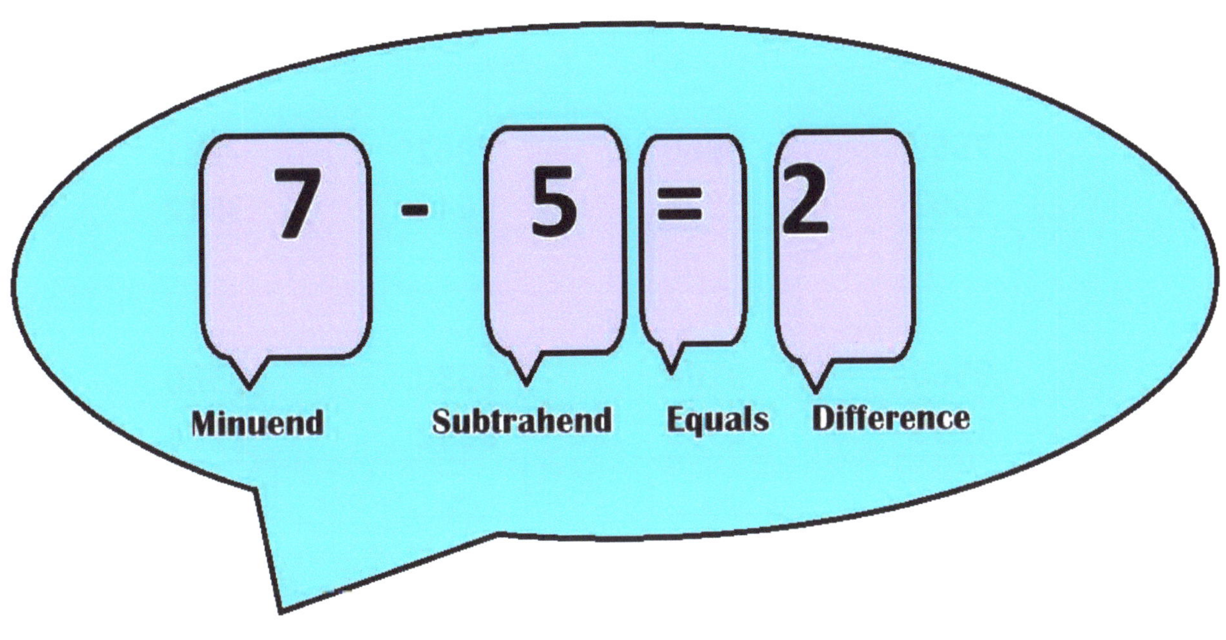

$$\begin{array}{r} 62125 \\ -\ 25641 \\ \hline 36484 \end{array}$$

$$\begin{array}{r} 56461 \\ -\ 23511 \\ \hline 32950 \end{array}$$

EXERCISE 2.0

6212 - 2564	5646 - 2351	6151 - 3500	3651 - 1224
6230 - 1562	6532 - 3325	6105 - 5642	6510 - 3512
7500 - 1257	7100 - 6251	7560 - 3564	7604 - 1256
7650 - 1562	7860 - 3325	7861 - 5642	7831 - 3512
9560 - 1250	9100 - 2565	9530 - 8892	9560 - 1224
4620 - 1562	4560 - 3325	4520 - 5642	4503 - 3512
6820 - 1562	6860 - 1235	6830 - 5642	6930 - 3512

EXERCISE 2.1

62125 - 25641 ☐	56461 - 23511 ☐	61512 - 35001 ☐	36512 - 12245 ☐
62300 - 15625 ☐	65320 - 33255 ☐	61055 - 56425 ☐	65100 - 35126 ☐
75001 - 12575 ☐	71005 - 62515 ☐	75600 - 35644 ☐	76044 - 12565 ☐
76501 - 15625 ☐	78600 - 33255 ☐	78611 - 56425 ☐	78311 - 35125 ☐
95600 - 12500 ☐	91005 - 25655 ☐	95300 - 88925 ☐	95600 - 12245 ☐
46200 - 15624 ☐	45600 - 33254 ☐	45200 - 56425 ☐	45030 - 35125 ☐
68200 - 15625 ☐	68600 - 12354 ☐	68300 - 56424 ☐	69300 - 35125 ☐

EXERCISE 2.2

62132	56400	61511	36533
- 25600	- 23500	- 35011	- 12233
62322	65311	61032	65133
- 15636	- 33266	- 56423	- 35155
75011	71032	75623	76066
- 12565	- 62532	- 35655	- 12522
76533	78611	78622	78322
- 15622	- 33277	- 56433	- 35230
95655	91033	95322	95351
- 12511	- 25655	- 88936	- 12235
46299	45633	45296	45056
- 15622	- 33255	- 56455	- 35123
68232	68636	68336	69392
- 15626	- 12335	- 56424	- 35111

EXERCISE 2.3

621325 - 256005	564005 - 235005	615115 - 350115	365334 - 122335
623225 - 156365	653115 - 332665	610325 - 564234	651335 - 351555
750115 - 125652	710322 - 625322	756235 - 356555	760665 - 125225
765332 - 156221	786112 - 332771	786222 - 564332	783225 - 352305
956554 - 125115	910335 - 256555	953225 - 889365	953515 - 122355
462995 - 156224	456336 - 332555	452965 - 564555	450565 - 351235

EXERCISE 2.4

621325	564005	615115	365334
- 265155	- 361551	- 266541	- 255115
☐	☐	☐	☐
623225	653115	610325	651335
- 215555	- 242565	- 235654	- 566100
☐	☐	☐	☐
750115	710322	756235	760665
- 235352	- 221111	- 255589	- 236541
☐	☐	☐	☐
765332	652531	786222	783225
- 361251	- 251551	- 562151	- 562151
☐	☐	☐	☐
956554	910335	953225	453351
- 365151	- 265255	- 365155	- 255155
☐	☐	☐	☐
831524	652152	892141	985641
- 156224	- 332555	- 564555	- 351235
☐	☐	☐	☐

EXERCISE 2.5

62132525	56400556	61511556	36533454
- 26515525	- 36155156	- 26654155	- 25511555
62322552	65311556	61032555	65133556
- 21555545	- 24256554	- 23565455	- 56610055
75011556	71032256	75623565	76066556
- 23535256	- 22111155	- 25558965	- 23654152
76533255	65253144	78622255	78322556
- 36125155	- 25155155	- 56215155	- 56215155
95655454	91033555	95322556	45335145
- 36515155	- 26525552	- 36515556	- 25515556
83152456	65215256	89214165	98564132
- 15622456	- 33255565	- 56455565	- 35123532

EXERCISE 2.6

62132525	56400556	61511556	36533454
- 45626545	- 35125645	- 23546812	- 36515425

62322552	65311556	61032555	65133556
- 32516515	- 32516451	- 45261251	- 12546584

75011556	71032256	75623565	76066556
- 56215621	- 35152651	- 36512541	- 36512541

76533255	65253144	78622255	78322556
- 32561254	- 32552412	- 23512541	- 12564251

95655454	91033555	95322556	45335145
- 65215213	- 32651251	- 32516524	- 12546251

83152456	94251245	89214165	98564132
- 32512451	- 35125142	- 32515412	- 32515421

3. MULTIPLICATION

Multiplication is a mathematical operation that represents the process of adding a number a certain number of times. It's often denoted by the multiplication symbol (×) or an asterisk (*).

Example

3 × 4

This means adding 3 together 4 times:

3 + 3 + 3 + 3 = 12

So, 3 × 4 = 12.

Here, we are going to learn the method and process of Multiplication with some illustration and examples.

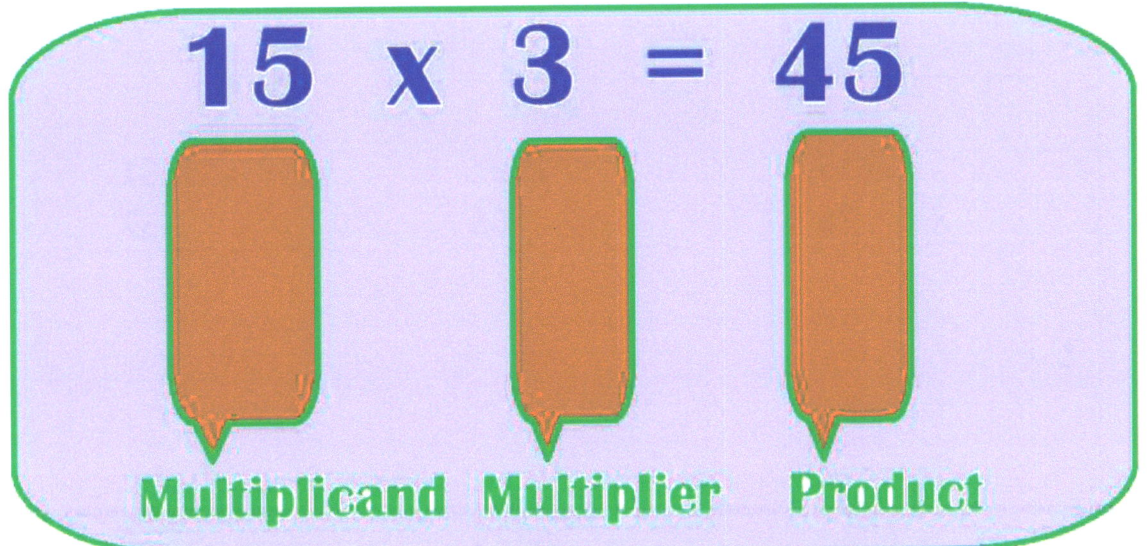


```
  2 5 6 1
    X 5 6
 15 3 6 6
12 8 0 5 X
14 3 4 1 6
```

```
 3 6 2 1
    X 1 3
1 0 8 6 3
3 6 2 1 X
4 7 0 7 3
```

EXERCISE 3.0

6212 x 12	4521 x 25	3651 x 65
3564 X 32	9265 x 32	5632 x 29
7215 X 21	7591 x 65	3691 x 52
6322 X 20	6325 x 23	5521 x 12

EXERCISE 3.1

6512 x 32	5521 x 20	3991 x 75
3004 X 52	9555 x 62	5602 x 29
4515 X 11	7661 x 25	3622 x 15
6356 X 15	8995 x 23	5881 x 12

EXERCISE 3.2

6512 X 132	5521 x 320	3991 x 175
3004 X 252	9555 x 262	5602 x 229
4515 X 511	7661 x 125	3622 x 615
6356 X 155	8995 x 253	5881 x 512

EXERCISE 3.3

4512 X 132	5781 x 320	5591 x 175
9904 X 252	9885 x 262	5502 x 229
4415 X 511	7881 x 125	6622 x 215
6666 X 565	8775 x 293	6681 x 312

EXERCISE 3.4

4562 X 932	5981 x 360	5191 x 105
9114 X 212	2085 x 262	5552 x 209
4665 X 501	7899 x 120	6682 x 285
6566 X 115	5875 x 293	6561 x 552

EXERCISE 3.5

4562 X 4932	5981 x 4360	5191 x 4105
9114 X 4212	2085 x 2262	5552 x 2209
4665 X 4501	7899 x 9120	6682 x 5285

EXERCISE 3.6

5562	5001	5111
X 4932	x 4360	x 4115

9554	2225	5662
X 1212	x 1262	x 2209

4115	7119	6552
X 4501	x 9120	x 5285

4. DIVISION

Division - Division is a mathematical operation that involves sharing or grouping a certain quantity into equal parts.

Illustration

Imagine you have 12 cookies that you want to share equally among 4 of your friends.

Step 1: Divide the cookies into 4 groups

Group 1: * (3 cookies)

Group 2: * (3 cookies)

Group 3: * (3 cookies)

Group 4: * (3 cookies)

Step 2: Check if each group has an equal number of cookies

Yes, each group has 3 cookies.

Result

12 ÷ 4 = 3

Each friend gets 3 cookies.

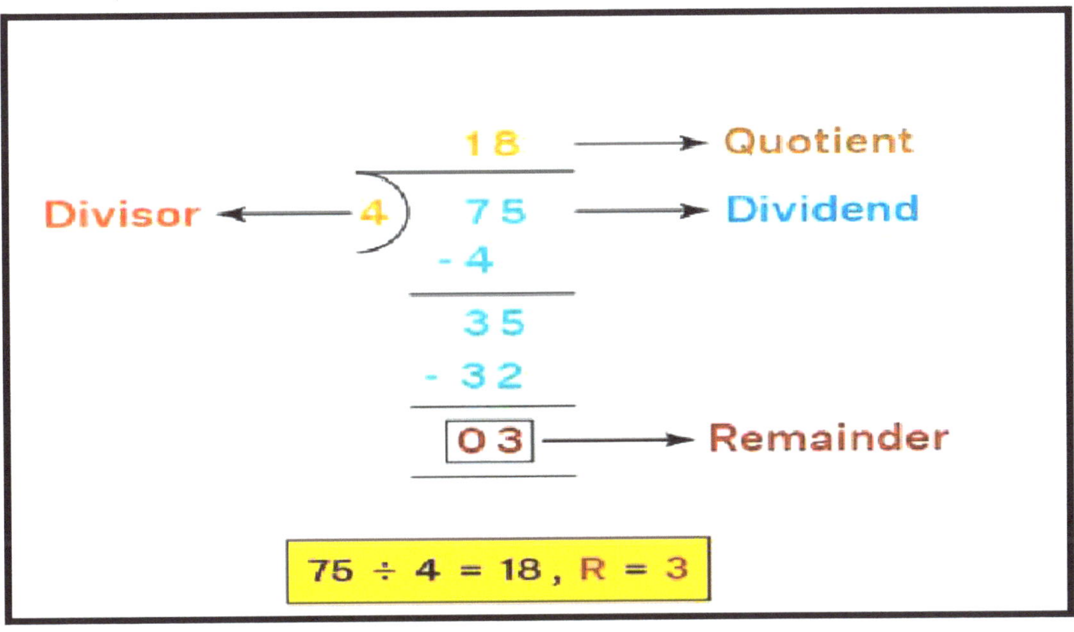

EXERCISE 4.0

14 ÷ 2 =	
24 ÷ 2 =	
18 ÷ 3 =	
16 ÷ 4 =	
12 ÷ 3 =	
14 ÷ 7 =	
34 ÷ 2 =	
32 ÷ 4 =	
60 ÷ 5 =	
40 ÷ 5 =	
45 ÷ 5 =	
55 ÷ 5 =	
62 ÷ 2 =	
46 ÷ 2 =	
48 ÷ 4 =	
56 ÷ 7 =	
66 ÷ 3 =	

50÷ 2 =	
40 ÷ 2 =	
30 ÷ 3 =	
40 ÷ 4 =	
21 ÷ 3 =	
49 ÷ 7 =	
92÷ 2 =	
55 ÷ 5 =	
70 ÷ 2 =	
63 ÷ 9 =	
56 ÷ 8=	
52 ÷ 2 =	
80 ÷ 2 =	
90 ÷ 2 =	
60 ÷ 4 =	
35 ÷ 7 =	
26 ÷ 2 =	

EXERCISE 4.1

28 ÷ 2 =	
24 ÷ 2 =	
20 ÷ 2 =	
44 ÷ 2 =	
36 ÷ 3 =	
49 ÷ 7 =	
32 ÷ 4 =	
36 ÷ 4 =	
45 ÷ 5 =	
40 ÷ 8 =	
98 ÷ 2 =	
96 ÷ 2 =	
94 ÷ 2 =	
99 ÷ 3 =	
90 ÷ 2 =	
88 ÷ 4 =	
86 ÷ 2 =	

21 ÷ 7 =	
42 ÷ 7 =	
36 ÷ 4 =	
75÷ 5 =	
85÷ 5 =	
64 ÷ 4 =	
78÷ 6 =	
70÷ 5 =	
68÷ 4 =	
57 ÷ 3 =	
80÷ 5=	
75 ÷ 3 =	
72 ÷ 8 =	
84 ÷ 7 =	
96 ÷ 8 =	
72 ÷ 4 =	
55 ÷ 5 =	

EXERCISE 4.2

128 ÷ 2 =		210 ÷ 7 =	
224 ÷ 2 =		420 ÷ 7 =	
202 ÷ 2 =		360 ÷ 4 =	
144 ÷ 2 =		175÷ 5 =	
336 ÷ 3 =		185÷ 5 =	
490 ÷ 7 =		164 ÷ 4 =	
332 ÷ 4 =		178÷ 2 =	
364 ÷ 4 =		705÷ 5 =	
455 ÷ 5 =		684÷ 4 =	
408 ÷ 8 =		570 ÷ 3 =	
198 ÷ 2 =		800÷ 5 =	

EXERCISE 4.2

128 ÷ 4 =	
224 ÷ 4 =	
210 ÷ 2 =	
144 ÷ 4 =	
336 ÷ 4 =	
490 ÷ 5 =	
336 ÷ 4 =	
564 ÷ 4 =	
855 ÷ 5 =	
448 ÷ 8 =	
198 ÷ 9 =	
964 ÷ 2 =	
940÷ 2 =	
996 ÷ 3 =	
908÷ 2 =	
888 ÷ 4 =	
812 ÷ 2 =	

280 ÷ 7 =	
560 ÷ 7 =	
480 ÷ 4 =	
275÷ 5 =	
485÷ 5 =	
264 ÷ 4 =	
540÷ 9 =	
755÷ 5 =	
984÷ 4 =	
760 ÷ 3 =	
850÷ 5 =	
750 ÷ 6 =	
960 ÷ 8 =	
840 ÷ 6 =	
960 ÷ 4 =	
880 ÷ 4 =	
555 ÷ 5 =	

EXERCISE 4.3

128 ÷ 16 =	
324 ÷ 12 =	
210 ÷ 10=	
144 ÷ 12 =	
336 ÷ 14 =	
490 ÷ 10 =	
336 ÷ 12 =	
560 ÷ 10 =	
850 ÷ 10 =	
108 ÷ 12 =	
104 ÷ 13 =	
117 ÷ 13 =	
112÷ 14 =	
126 ÷ 14 =	
105 ÷ 15 =	
135 ÷ 15 =	
128 ÷ 16 =	

280 ÷ 14 =	
560 ÷ 14 =	
480 ÷ 12 =	
126÷ 18 =	
480÷ 10 =	
133 ÷ 19 =	
114÷ 19 =	
750÷ 15 =	
980÷ 10 =	
760 ÷ 10 =	
850÷ 17 =	
750 ÷ 10 =	
960 ÷ 10 =	
840 ÷ 10 =	
660 ÷ 10 =	
880 ÷ 11 =	
440 ÷ 11=	

5. METRIC SYSTEM

In our daily life, we come across situation where we want to know about the lenth of a particular object or where we have measured a particular thing. Suppose, you are telling your friend about the distance to your house from the Market, you would express this distance in Kilometers.

When we are going to market to purchase any things for daily uses, this will also measure in term of kilo, gram, meters, and liters. For example, we are going to purchase rice then, we have to told to the shopkeeper about please give me one kilograms of rice, and when we want to purchase Milk then this is measure in term of Litre.

FOLLOWING IS THE MEASUREMENT OF UNITS.

- FOR MEASUREMENT OF DISTANCE & LENGTH.
- 10 MILLIMETRES = 1 CENTIMETRE
- 100 MILLIMETRES = 1 DECIMETRE
- 1000 MILLILITRES = 1 METRE
- 100 CENTIMETRES = 1 METRE
- 10 DECIMETRES = 1 METRE

1000 METRE = 1 KILOMETRES

- FOR MEASUREMENT OF MASS OR WEIGHTS.
- 10 MILLIGRAM = 1 CENTIGRAM
- 100 MILLIGRAM = 1 DECIGRAM
- 1000 MILLIGRAM = 1 GRAM
- 100 CENTIGRAM = 1 GRAM

1000 GRAMS = 1 KILOGRAMS

5mm + 5mm = 1CM

500cm +500cm = 1Metre

500m+500m= 1KM

EXERCISE 5.0

Example 1- Convert the following into centimetre (cm):-

(a) 7m (b) 6m 40 cm (1m=100cm)

Solution :- (a) 7m = 7 x 100cm = 700cm

(b) 6m 40 cm = 6 x 100cm + 40cm = 600cm + 40 cm = 640cm

1. 7m 15cm =
2. 8m 10cm=
3. 15m 60cm =
4. 14m 40cm=
5. 12m 40cm=
6. 10m 40cm=
7. 18m 60cm=
8. 13m 60cm=
9. 15m 45cm=
10. 13m 66cm =
11. 21m 20cm=
12. 20m 25cm=
13. 12m 30cm=
14. 14m 45cm=
15. 18m 10cm=
16. 18m 58cm=
17. 16m 65cm=
18. 45m 40cm=
19. 30m 40cm=
20. 20m 60cm=
21. 23m 56cm=

EXERCISE 5.1

Example 1- Convert the following into metres (m):-

(c) 7km (b) 6km 40 m (1km=1000m)

Solution :- (a) 7km = 7 x 1000m = 7000m

(d) 6km 40 m = 6 x 1000m + 40m = 6000m + 40 m = 6040m

1. 7km 15m =
2. 8km 10m=
3. 1km 6m =
4. 14km 40m=
5. 12km 40m=
6. 10km 400m=
7. 18km 600m=
8. 13km 60m=
9. 15km 45m=
10. 13km 66m =
11. 21km 2m=
12. 20km 25m=
13. 12km 3m=
14. 14km 400m=
15. 16km 100m=
16. 15km 580m=
17. 16km 650m=
18. 45km 400m=
19. 30km 40m=
20. 20km 600m=
21. 23km 60m=

6. MONEY

In our daily life , we need purchase goods from the market. In exchange, we pay money to the shopkeeper in the form of coins and notes.

The type of money particularly used in any country is called its currency.

In Our Conutry, we used the units of Indian currency in the form of rupees and paisa.

At present time we have the following Indian coins as below-

Rs.2

Rs.5

Rs.1

Rs.10

A part from coins, we have the currency in notes form. we use these notes in our life, all these notes are in the form of denomination of one-rupee, two-rupee, five-rupee, ten-rupee, twenty-rupee, fifty-rupee, hundred-rupee, two-hundred, five hundred-rupee and two-thousand rupee.

EXERCISE 6.0

Following is some images of Indian currency; the task is identified the value of the currency and coins.

EXERCISE 6.1

Following is some images of Indian currency; the task is identified the value of the currency and coins and solve the sums-

₹2000
+
₹2000
+
₹200
=
₹500
+
₹2000
+
₹50
=
₹50
+
₹50
+
₹200
=
₹500
+
₹500
+
₹50
=

7. CLOCK AND TIME

Clock- A Clock is used to measure time in our daily life, the clock has two hands- the short hand is called the hours hands and the long hand is called the minute hand.

Time- Time is a measure of the duration between events, allowing us to understand the sequence and progression of events. It's a fundamental concept that helps us organize our daily lives.

AM and PM are abbreviations used to divide the day into two 12-hour periods:

AM (Ante Meridiem)

- Means "before midday" or "morning"
- Refers to the period from midnight to 11:59 in the morning
- Examples: 6:00 AM, 9:00 AM

PM (Post Meridiem)

- Means "after midday" or "afternoon/evening"
- Refers to the period from 12:00 noon to 11:59 at night
- Examples: 1:00 PM, 8:00 PM

Examples

- 7:00 AM: 7 o'clock in the morning
- 7:00 PM: 7 o'clock in the evening

TIME- 3:00

TIME-4:00

TIME- 5: 15

EXERCISE 7.0

In the chapter, we are going to solve of the clock and time question using of Hours Hand and Minutes Hand.

EXERCISE 7.1

In thiss exercise, wc arc going to solvc thc Clock qucstion and find out thc time which is indicate in the clock.

www.ingramcontent.com/pod-product-compliance
Ingram Content Group UK Ltd.
Pitfield, Milton Keynes, MK11 3LW, UK
UKHW060107300726
14090UKWH00003B/394

* 9 7 9 8 8 9 9 0 6 3 5 5 8 *